Raising Goats

Some Essentials

By Cheryl K. Smith

karmadillo Press

karmadillo Press
22705 Hwy 36
Cheshire OR 97419
USA
(541) 998-6081
www.karmadillo.com

Author: Cheryl K. Smith
Cover Design: Annette Wilson

ISBN: 978-1-7335274-2-2

Also from karmadillo Press and
available at Amazon.com:

Goat Health Care, 2nd Edition

Goat Behavior, by Tamsin Cooper

*Best of Ruminations Goat Milk
and Cheese Recipes*

Table of Contents

Some Goat Terminology

Doe—a female goat. Sometimes called a nanny. Young females are called doelings

Buck—a male goat. Sometimes called a billy. Young males are called bucklings.

Wether—a castrated male goat. They make the best pets.

Disbudding—removal of the horn buds when the goat is still very young.

Polled—naturally hornless.

Scurs—parts of horn that have grown back after disbudding.

Estrus, or heat—the time when a female is capable of breeding. This usually occurs every 21 days for 12 to 36 hours.

Rut—the season when bucks are ready to breed. They act aggressive and pee on themselves.

Buck rag—a cloth that is rubbed on a buck and given to a doe to smell to determine whether she is in heat.

Kidding—giving birth.

Freshening—coming into milk.

Ruminations—chewing cud, or the sound made when a goat is digesting food—like a growling stomach

Basic Physiological Data

Normal Temperature	101.5–105.0 °F
Pulse	70–80 beats/min
Respirations	12–15 breaths/min
Ruminations	1–1.5 minute
Gestation	145–155 days

Tip:

When you get new goats, check these vital signs several times and record them on a calendar or goat health record. You can use this information to determine whether the goat is sick when it is acting "off" and will help a veterinarian better understand what is going on.

Getting Your Goats

Goats are intelligent, clean and curious animals, despite some of the myths about them. They rival the pig and the dog in terms of intelligence. At times this intelligence, combined with their high energy and curiosity, can cause problems. That is why it is important to prepare before you get goats.

What kind of goat should you get?

Miniature dairy goats are great for urban settings or a small acreage. Mini dairy goat breeds include the Nigerian dwarf and various crosses of standard dairy goat breeds, such as Oberian, Mini Nubian, Mini Alpine, or MiniMancha. Of course, if you want fewer goats, but a better milk supply, get full-sized goats.

Maybe you are interest in fiber and fiber arts. There are a variety of goats that will fill this need. Angoras, Cashmeres, Pygoras, or Nigoras can all be good choices.

Any of the above goats can be used for meat, but if it is meat goats you are interested in, there are Boers, and many other crossbreeds.

Do you want to show your goats? Are you interested in breeding? How much space do you have?

If you want a milker, you will obviously need to get a doe. Nigerian dwarves generally milk 2–4 lbs per day and the other minis can milk even more—often double that amount. If you want more milk, but fewer goats, consider mini Alpines or mini Toggs. Their full-size relatives are known for high milk production. A mini Nubian will give you more milk than a Nigerian, but with comparable butter fat and protein, which means more cheese.

MiniManchas, like their full size relatives, are cute and extremely friendly and gentle. Or you may want to choose an Oberian because you like the markings, temperament and milk production.

If you are not interested in milking, but just want a pet or brush eater, a wether (castrated male) is your best bet. They are inexpensive—priced between $50 and $100, depending on where you live—and, in my opinion, have the best personalities. They never go into heat, they don't stink, and they are very loving.

How many goats should you get?

This may seem like a strange question, unless you know goats. They are herd animals, so having just one goat is usually a bad idea. It will get lonely and bored, unless paired with another animal like a horse or a sheep. Goats are not generally good matches with dogs, in terms of companionship. So assume that you are going to purchase two goats.

Because I wanted to milk and make dairy products, I first opted for two doelings. Some people go with just one doe and a wether to keep her company. Getting a buck and a doe for a starter herd is not a good idea, because you are then very limited on genetics and breeding options. I leased a buck for the season my first year. This gave the buck an opportunity to live with two nice girls for a few months; a much better deal than living with a bunch of boys. Before you impulsively buy a buck to breed with your does, seek out local goat owners and talk to them about leasing a buck.

What is needed for a goat?

First, you need sturdy *fencing*. Goats will eat your rose and lilac bushes, and any garden you want to plant. Because some landscaping plants may be poisonous to them, you will need to either keep the goats away from them or keep them away from the goats. I like welded wire fencing for mini dairy goats. The kids can get out of field fencing and cattle panels.

Second, your goats should have a *shelter,* which can vary depending on your location and protection from predators. You can get away with several small dog houses or dog igloos for mini goats that are in a safe area. I have had up to three goats sharing the igloo I put in their pen. They will also need straw or wood shavings for bedding. My goats prefer wheat straw.

If you are not using a shelter that can be closed up at night, you need to consider getting a livestock guardian dog or another animal such as a llama or donkey for protection. That is a subject for another book.

You need *water buckets* because goats require lots of fresh, clean water. If you don't have hot water in the barn, a heated bucket may be a good investment for the winter.

Feeding bowls are another requirement. Bucks or wethers don't usually need grain, and can even develop urinary stones from it, while kids and does that are producing milk or pregnant should be fed some grain. You also will use a feeding bowl for treats like kelp, peanuts, nutritional yeast, or vegetables.

In addition, for all types of goats, you need a *hay feeder*. Goats are notorious hay wasters and once hay is on the floor, it becomes bedding, so you don't want to start with it there. What hay you feed will vary with what's available, but try to find good quality hay. I feed grass/alfalfa mix whenever possible, along with cheaper, local hay.

Finally, a goat *mineral block* or *loose minerals* are essential. You will want to buy or make a container to hold the minerals and keep kids off of them. Minerals ensure that your goats grow well and they supplement the feed and forage.

In terms of equipment, if you only have a few goats and are not going to have kids born, you need very little. I do have to mention that buying or building a *milk stand* (also called a stanchion) is one of the best investments you can make.

Goat Social Behavior

You can't have a goat herd for long without becoming something of an amateur goat sociologist. Sociology is the study of groups, and goats are an excellent group to study. (Actually, the correct term for study of animal behavior is ethology.)

Learning about how goats interact with each other is as important as learning how to feed and care for them as individuals, whether they are pets or producers.

In their natural state, goats tend to establish a dominance pattern within their herd. Age and sex are determining factors in caprine dominance patterns. (Horns can also play an important role, but since most dairy goats are disbudded or hornless, this article focuses on the other factors.) When does and bucks are kept separate, age becomes the most important factor in dominance, although as any goat owner can attest, personality plays a definitive role.

The herd has a herd queen, who leads the other goats to the best grazing areas (although she may try to keep the best for herself, her daughters and her chosen friends). My observations have indicated that this is not always the case, but when she does go, everyone follows. Removing the herd queen can confuse the other goats.

In the wild, goat herds also have a top buck, who will take up the rear of the herd to protect against predators.

The dominance pattern among goats is established with the original herd. It is likely to continue until the lead doe or buck dies, or in some cases, gets old and/or infirm. A discussion on one of the e-mail goat lists dealt with a herd queen being challenged for her position shortly after she and the challenger gave birth. After several separations, the two gave up their battle, with the original herd queen giving up her position to the challenger. I have witnessed this type of challenging behavior after kidding, especially if the herd queen has been removed from the herd and placed in a kidding pen for a period of time. While she is out of the herd, the next goats in line have been staking out their territory, only to discover (usually) that the situation was temporary.

Goats establish dominance through head-butting and other aggressive acts, such as biting. The dominant goat will get to eat first, drink first, do everything first!

In the wild, or when they are allowed to run together, the top buck will dominate the herd, with the herd queen second. When separated, the herd queen will dominate the group of does and kids. The herd queen will always be dominant over her doe kids, even after many years.

I've observed in my herd that the doe kids of my herd queen automatically move up near the top of the herd in dominance. They are obviously used to special treatment as, from their birth, their mother would run off any goat that got too close (other than her older daughters).

According to a fact sheet written by E.A.B. Oltenacu and Tatiana Stanton for the New York State 4-H Dairy Goat Project, in the wild, when the herd queen finds a poisonous or inedible plant, she will sniff it and then snort and show obvious dislike for it. All the goats in the herd will then smell the plant, one at a time, using the scent to identify that plant. Then after all the goats have memorized the plant's scent, the top buck will trample the plant.

Some natural goat behaviors may not occur in domesticated herds, because of different conditions that affect how they act. These include all the ways in which humans interact with their goats. For example, when humans feed the goats they become associated with the herd queen and may have problems getting the goats out to browse. My goats are so conditioned by my feeding, that when I drive up in my car, or walk from the house to the barn, they all run in from the pasture.

Goat watching is fun and always brings the new and unexpected, as well as the old and expected. Now take some time to go out and observe your goats' behaviors.

Keeping Your Goats Healthy

Start with Healthy Goats

Avoid livestock auctions, unless you want to rescue goats. Sometimes people take their goats to the auction because there is something wrong with them—including an incurable disease such as caprine arthritis encephalitis virus (CAEV).

Go to the farm where you are getting them, if possible, and observe the herd. This will give you a chance to see how the goats are living, whether others seem healthy, and it will also give you ideas for how to set up your own area. Look for:

- ✓ Body condition

- ✓ Ruminating

- ✓ Normal poop

- ✓ Condition of coats

- ✓ Cleanliness and space in accommodations

- ✓ Clear eyes and nose

Ask questions. Ask the goatkeeper about any prior or current problems that may have occurred in the herd. Learn about their feeding program. Find out why they are selling the goats. Other questions to ask include:

- ✓ What diseases do you test for?

- ✓ Do you vaccinate? If so, for what?

✓ Is there any history of abortion in the herd. If so, do you know what caused it?

✓ Will you give me names and contact information of people you previously sold goats to?

Get it in writing. You can use the contract in this book as a basis. It is best to have the agreement in writing. Reputable breeders will often replace goats that die soon after purchase. If you are buying an animal for breeding, the contract should have a provision that you will get a replacement animal if the one you purchase is not fertile. Contracts lay out what the buyer and seller are agreeing to, to avoid misunderstandings.

Quarantine new goats. If new goats are being added to an existing herd, it is best to quarantine those brought into the herd for at least 30 days. This provides a chance to watch them for problems, do any necessary testing or treatments, and interact with them. During this time, test for parasites and deworm, if needed. Get a baseline temperature and vital signs.

Find Resources

Veterinarian. Find a veterinarian and schedule an appointment for one or more of your goats to establish a relationship. Goat veterinarians can be hard to find and are in demand. If you cannot find a goat vet, try to find one who is willing to learn about goats and do what you can to educate her. If you already have a relationship, they will be more likely to come to your aid when you have a goat emergency. A relationship is also required for the prescription of medications that are not over-the counter or are extralabel.

Mentor. Find a person with experience in goats. It may be the person who sold you goats, a member of a local goat club, or even a neighbor who has goats. I usually offer to answer questions that come up for anyone I have sold goats to because I learned about goats from the first people who sold them to me

References. There are a lot of good goat books out now, including *Goat Health Care (*www.goathealthcare.com) and *Raising Goats for Dummies,* which I wrote. Read reviews on books on Amazon.com or another bookseller before buying them, though. There are as many good ones as bad ones out there. *Goat Journal* is a magazine that you can subscribe to, as well.

There are some good websites, as well, but with the same caveat: be careful with advice. There is good and bad advice being proffered. The Maryland Small Ruminant Page (www.sheepandgoat.com), International Veterinary Information Service (www.ivis.org), and the American Consortium for Small Ruminant Parasite Control (www.wormx.info) are all outstanding. There are many more.

Provide Proper Living Accommodations

Shelter. Goats need secure and comfortable shelter. Most goats hate the rain and for good health, they need at least a shelter with protection from the wind, too. Wheat straw or wood shavings make good bedding and, because goats like to be up, they will enjoy a bench or other elevated structure.

Fencing. Make sure your fencing is high enough. I recommend avoiding electric fencing, for the rare occasion that a goat will get its head stuck and be injured or killed. Cattle panels will work for larger goats, but a secondary fence at the bottom is often necessary to keep kids in. If you have an already-existing fence, check around the edge for safety hazard.

Cleanliness. Keep their bedding clean. Try to avoid letting the mud build up. Make sure they have a dry area. Especially in the winter, this does not mean cleaning every day. Instead, you can add dry straw to the bedding areas. This buildup, while creating a bigger job when you finally much, will help keep goats warm in cold weather, as the material below composts.

Do not overcrowd. Like with any other animal, including humans, overcrowding can lead to the spread of disease. Also make sure that kids and other vulnerable goats have a place to get away from others that may want to bully them. This will make for a less stressful environment, and better health in goats.

Pest Control. Flies, rats, mice, and parasites can all be problems in barn living. Of course, cleanliness will go a long way in keeping them at bay.

Fly control can be especially challenging and everyone seems to have their favorite method for preventing them or killing them when they are at their worst in the summer months. My favorite method (after trying almost everything over the years) is Mr. Sticky fly reels. These are like horizontal fly strips, without the poison. The reels are installed and once the exposed strip is covered in flies, they can be reeled in to expose more sticky surface.

Know Your Goats

Regularly observe. At least twice a day, when you go in to do routine feeding, make sure to look at each of your goats. You will learn the temperament of each one, so if a goat who is normally friendly hangs back, go closer to check him or her out. Is the tail down? Does it have diarrhea? Is it hunched down? Shivering? Is the goat not interested in food? It is time to take a closer look and get your hands on that goat.

Take a temperature. This is where the baseline temperature comes in handy. Take the temperature of a goat you have observed not acting normal. Compare that to the baseline temperature (taken twice when you first get the goat). Is it high or low? It may be normal, but write it down and then continue checking the goat and keeping an eye on it for the next few days.

Listen for ruminations and watch for cud. You can hear the sound of the goat's digestive system by simply putting your ear on its left side. These sound like a stomach growling, and in a healthy goat they are loud and fairly frequent. This tells you that the goat's digestive system is working.

Another way to tell that the digestive system is working is to watch for cud-chewing. Goats eat and swallow their food and it goes into the first part of the stomach (which has four parts) and mixes with bacteria and digestive juices. Then the partially digested material (bolus) comes back up and the goat chews it for further digestion. This process is repeated throughout the day and eventually the bolus is passed on to the rest of the stomach for further digestion.

A good way to check for herd health is to look around to see whether your goats are chewing their cud. Normally about 2/3 at a time will be doing so.

Check body condition. Goats should not be fat or thin, but to be healthy they should be on the thinner side. Fat goats not only have fat on the outside, but it can also build up between internal organs. Read about body conditioning and practice feeling your goats, aiming for a condition of 2— which means you can feel their spine and ribs and they have a limited fat pad on the front of their body. There is lots of information on the Internet about body condition.

Feed Them Right

Feed. Provide high quality hay and/or browse. Because of their digestive system, goats need plenty of roughage. Some people feed chaffhaye, while others supplement with it. Chaffhaye is a fermented alfalfa that comes in a bag— which makes storage easier.

Goats are browsers, not grazers. Although they will eat grass, they prefer woody shrubs. Learn about some of the plants that can have medicinal effects. For example, fir branches and oak leaves can help control parasites. Moderation is the key and, if they are not starving, you will see your goats eating a little of this and a little of that in the pasture or woods.

Don't overfeed or underfeed. Learn about body condition scoring to determine whether your goats are over- or underfed. Don't give grain to dairy goats, unless they are pregnant or lactating. It is not necessary. Wethers are especially at risk, because they are more prone to urinary calculi.

Minerals. Minerals are essential. They come in blocks and loose. Ask other goat owners or your vet about the best mineral. Do not give sheep minerals because they don't have enough copper for goats. (Even more important, do not give goat minerals to sheep, because than can get copper toxicity.)

Fresh, clean water. Goats like their water clean. In the winter, they even prefer it warm or hot (not too hot). I have found that using smaller buckets and changing it regularly works best for me. In order to stay healthy, and to help prevent urinary calculi in wethers, water must always be available.

Other supplements. A variety of supplements may be provided to goats to help keep them healthy, although some can be expensive. I recommend kelp, nutritional yeast, sunflower seeds. To help put weight on a skinny goat, beet pulp is a good addition to the diet. Some people use free choice baking soda to balance the rumen. Peanuts in the shell, carrots, or fruit are good treats. (See the chapter on goat gardens for ideas.) Goats are also known to have copper deficiency, but they can develop copper toxicity if given too much, so talk with your vet or read up on this before adding it.

Keep Them Safe

Never tether goats unless you can be there to watch and supervise. They can get tangled up, hang themselves, lose access to shelter or water, or become the bait for a predator. A tethered goat cannot get away from a dog (the most common predator), a cougar or a coyote. In addition, if one goat gets tangled up, unable to move, another nearby goat may find it is the perfect opportunity to bully and bash.

16

Protect from predators. Besides not tethering them where predators can reach, goats particularly need protection at night. Fencing is important, but living where wildlife also co-exist requires that the animals be safely secured at night, or have a protector. If you cannot put them in a closed barn, make sure to have a livestock guardian dog (LGD), a gelded llama, or a donkey. LGDs, such as Great Pyrenees, Maremmas, or Anatolian Shepherds are well worth the money they cost, if you have a medium-sized to large herd.

Minimize or remove hazards. Before getting goats, walk your property to ensure that hazards are removed. These include things like defective fencing or barbed wire that can cause injury, poison plants, plastic, or sharp metal. Make sure that any goat feed, other than hay, is stored in a bin or garbage can that they cannot access. I prefer metal to plastic garbage cans, because rats can chew through plastic.

Biosecurity. If you have a lot of visitors, especially people from other farms, have them wash their hands, or keep them separated from the goats. This may not always be feasible, and if you are concerned about what they might be bringing in on their shoes, provide slip on plastic covers or even separate boots for them.

Keep up on Routine Care

Vaccinations. Most veterinarians recommend that goats be vaccinated with CDT, at a minimum. Other vaccines for various diseases are available, so find out from your vet what is recommended. Make sure to keep up on annual vaccinations.

Deworming/fecals. A fecal exam should be done before deworming a goat. You can collect feces and mail them to a lab, or get a microscope and do it yourself. Images can be found online. It is good to do a fecal at least annually and when your goat has a problem such as diarrhea that may be caused by parasites.

Hoof trimming. A goat's hooves need to be trimmed regularly, from every two months to longer, depending on the living conditions, and whether there are rocks or gravel that wear the hooves down. This is a fairly simple task that will pay off in the long run, because having long, painful, or deformed hooves can lead to problems with the legs, hips, and other parts of the body. They are also more prone to hold mud, which can lead to foot rot.

Side benefits of frequent hoof trimming include keeping the goat tamer and providing an opportunity to fully check the goat over to ensure good health and body condition.

Recordkeeping. Keep records of hoof trimming, vaccinations, preventive care, and any illness and how it was treated. You can also use records to keep track of when goats are in heat or when they kidded, how many kids they had, how they did at or after the birth, and even days of gestation, if you hand breed or otherwise observe the breeding.

Testing. Unless you have a closed herd (no new goats in, no showing, no exposure to other goats or sheep), you will need to have your goats tested. The most common test is for CAE virus, a disease like human HIV that can be spread through fluids. Other common tests are brucellosis (find out if your state is certified brucellosis-free), caseous lymphadenitis, and Johne's disease.

Goats and Gardening

Goats and gardening go hand in hand. Goats benefit the garden with their manure, which helps to produce healthier plants and greater crop yields. The manure is in pellets, so it can be applied directly and not burn plants like chicken or cow manure does. It does not have a strong smell, either, which is a plus.

The nutrients in the manure aid in plant growth and the urine that is contained in the bedding being used adds to the nitrogen content, which makes it a more effective fertilizer once composted. Working it into the soil improves the texture, aiding in more efficient use of water and allowing more oxygen to get to the roots of the plants.

Another reason for composting goat manure is to destroy weed seeds that may have survived their trip through the goats' digestive systems, and to kill any pathogens that may be contained in the manure

Gardens can benefit goats, if you plant vegetables that are good for goats and which they enjoy eating. *As you plan or plant a garden in the spring, why not add some plants with goats in mind?*

Sugar snap peas are usually the first plant to go in each spring. Make sure to plant away from the fence line or the goats will make quick work of them. But they make great treats for the goats—both pods and leaves—when they have quit producing and are ready to be taken down. I usually pick some for the bucks when getting my own, during that time when they are exploding with pea pods.

Sugar snap (or snow) peas are high in fiber, vitamin A, vitamin C, vitamin K, thiamin, and folate (a B vitamin), iron, and contain several minerals, including magnesium, phosphorus, iron and potassium.

One of my goats' favorites is **carrots**. I always plant enough to get me through the winter, if possible. When I pick them, I often remove the top greens and feed them directly to the goats. These greens are high in manganese, which is good for bones and skin and helps to control blood sugar.

I also cut up carrots and give them to the goats in a big bowl. The roots (carrots) are full of vitamins C, B, B_2, and A. One cup of raw carrots has over 1200 IU of vitamin A. They are a good source of fiber and antioxidants, too. One study (in humans) found decrease in cardiovascular disease correlated with carrot intake. Carrots also contain the minerals molybdenum, potassium, copper and phosphorus.

Beets are another great root vegetable that are nutritious for goats. As with carrots, goats can eat both greens and the root. The beet should be cut into smaller pieces to avoid choking.

Beets are high in manganese, vitamin A, folate (a B vitamin) and vitamin B-6, vitamin C, thiamine, potassium, riboflavin, niacin and pantothenic acid.

On the downside, beets contain oxalates, which if eaten if a large quantity can lead to urinary stones—a problem found more commonly in wethers—and oxalate toxicity, which can cause restlessness, muscle tremors, weakness, torticollis (a twisting of the head to one side, with painful muscle spasms) and ultimately death.

Another plant that naturally goes with goats is the **sunflower**. While growing sunflowers and harvesting all the seeds for goats would probably not be worth the labor if you were trying to replace the black oil sunflower seeds (BOSS) that many add to their goats' grain ration, they can be a nice supplement. The whole plant is edible, too, including stalk and leaves. And the seeds don't have to be hardened for the goats to like them. If you have plenty of sunflowers, just break the head off of one or more and break into pieces for the goats or remove a few leaves for a snack.

In terms of vitamins, sunflower seeds are loaded with vitamin E and also contain vitamins A, B, and D. They are high in protein and contain the minerals iron, zinc and magnesium. And goats love them!

Greens—ranging from lettuce to spinach to parsley and kale, are fast and easy to grow and provide a nutritious supplement for goats. These all provide dietary fiber to goats, and their nutritional values depend on the type of green. Iceberg lettuce has less nutritional value, while kale and spinach are better. They contain varying amounts of various vitamins and minerals, as well as oxalates. High-oxalate greens include kale, spinach and swiss chard, so should not be fed in large quantities.

Pumpkins and squash can produce a nice autumn snack for goats and will be available after some of the prior plants are done producing. Goats can eat both the fruit and the seeds. The seeds are a fantastic source of both zinc and vitamin E, as well as a diverse array of antioxidants. Besides being a good source of protein (some are up to 35% protein), pumpkin and other squash seeds contain the

21

minerals phosphorus, magnesium, manganese, iron and copper, and lots of B-complex vitamins.

Many types of squash will keep over the winter, so they can be doled out as treats during cold weather. They are also full of beta carotene and vitamin A, supporting good vision as well as aiding immune function and assisting with growth of bone and connective tissue.

Folklore has it that pumpkin seeds are an effective dewormer. A small study in Delaware showed a small decrease in fecal egg count in goats given pumpkin seed and some studies have shown similar effects in humans, but more study needs to be done. Relying on pumpkin seeds alone for preventing internal parasites is a mistake, unless you are also doing routine fecal exams to identify problems.

To feed squash or pumpkin to a goat, simply cut it in small pieces and serve in a bowl or break the squash open and put in the field. Most goats will be excited about it, but like with other foods, some may not be interested.

When adding any of these garden vegetables to a goat's diet, start feeding in small quantities—in addition to their normal diet of hay and any browse they eat—to avoid digestive upset. Rather than providing all of it at once, pick a little each day and share with your goats for an afternoon snack or at regular feeding times. They will love you for it and their health will show it.

Going Organic with Goats

Since the 1990s, the U.S. has seen a resurgence of interest in growing and eating organic foods. More people have become interested not only in organic fruits and vegetables, but in organic dairy products and meat.

According to the Organic Trade Association, a North American trade organization for the organic industry, in 2010 about three-fourths of Baby Boomers polled indicated that they purchase organic or natural foods. Yet, despite an annual growth rate of about 20% a year, the land dedicated to organic products is still small compared to that for foods grown by "modern" methods. Partly because of the limited supply, but also because of growing concern over genetically modified crops, exposure to pesticides, out-of-control use of antibiotics and other chemicals in food animals, and the perception that organic is better for health, people are willing to pay a higher price for certified organic food.

Besides providing organic meat, fiber, milk and other dairy products to meet growing market demand, raising goats organically has another big benefit: it promotes sustainability. Organic manure produced by the goats can be used as fertilizer (and become another product to sell); less soil is lost to erosion with organic processes; less pollution (particularly to groundwater) is created and less fossil fuel is used because chemical fertilizers have been eliminated.

This article covers the requirements for raising goats that can be organically certified, some of the difficulties in meeting these requirements and some alternatives that may serve the same purposes.

So, just what is considered organic goat milk or meat? **Organically certified** goat milk or meat is from goats that were raised according to the standards of the National Organic Program and certified by an accredited state or private agency. (See sidebar)

Organic Milk and Meat Requirements

- Produced by farmers who emphasize the use of renewable resources and the conservation of soil and water to enhance environmental quality for future generations.
- From animals that are given no antibiotics or growth hormones
- Produced without using most conventional pesticides; fertilizers made with synthetic ingredients or sewage sludge; bioengineering; or ionizing radiation
- Farm where produced is inspected and certified by a government agent to make sure the farmer is following all the rules necessary to meet USDA organic standards
- Companies that handle or process the meat or milk before it gets to the local supermarket or restaurant must be certified

Source: USDA Consumer Brochure: *Organic Food Standards and Labels: The Facts*

Organic is not the same as **natural,** which refers to products that are minimally processed with no artificial ingredients, coloring agents or chemicals. Organic is also not the same as **pasture-raised or grass-fed**, although this is a component in the organic requirements. Just this year, the US Department of Agriculture (USDA) further restricted the definition of organic milk and meat to require that it come from livestock that graze on pasture for at a third of the year, which get 30% of their feed from grazing. Previously they only had to have "access to pasture."

Switching to raising your goats organically may not be easy. There are a variety of requirements that some may find hard to meet, depending on their farm and the local conditions.

In addition to the federal organic standards, some states may have more restrictive requirements. Farmers need to check with their state Agriculture Departments to find out what, if any, requirements the state will impose before getting too far into development of a plan to go organic with their goats.

Living Conditions. The requirement that goats be on pasture at least one-third of each year entails more than just putting them in just any outdoor or indoor area for the required amount of time. The standard also requires that they have shade, shelter, exercise space, fresh air and direct sunlight.

Shelter must be provided in a way that promotes the healthy and natural behaviors and maintenance of the goats. This includes ensuring that they are safe (hazards are minimized to prevent injury), have an opportunity to exercise, are protected from severe temperatures, have adequate ventilation and have appropriate bedding (clean and dry).

A few exceptions allow temporary total confinement at times, including bad weather, health and safety issues and risk to soil or water quality, but they cannot be totally confined for all or even a majority of their lives.

Pasture fencing is also covered by the Organic Standard. Using treated wood is prohibited in organic production, where it may come into contact with soil crops or the livestock. While most older fencing may be grandfathered in, farmers who are starting with a new farm need to be aware of this prohibition and avoid using treated wood in fence construction.

Pasture. In order to be certified organic, goats must be raised on pasture that is certified organic. This requires the use of no pesticides, herbicides, chemical fertilizers or any other restricted materials. Unless the pasture land is already certified, obtaining initial certification will take time. In order to receive initial certification for pasture land, a farmer must be able to show that no prohibited substance has been applied for 36 months prior to full certification. Farmers with recently purchased property may have a harder time getting certified unless the prior owner is willing to assist with proving that no prohibited substances were applied.

Also important is the need to avoid overstocking of goats—which will lead to overgrazed pasture land. Overgrazed pasture, or pasture otherwise lacking vegetation, is not considered pasture under the certification standards any more than a feedlot is.

If a pasture abuts a roadway, where a county or other municipality may spray the roadside for weeds, put up signs indicating that the property is an organic farm and spraying is prohibited. Contact the municipality as well to make sure that you supply them with any required documentation.

In organic management of goats, pasture rotation may be even more important, because use of most chemical dewormers is prohibited. Pasture rotation discourages parasite overpopulation, especially in warm, wet regions, as well as discouraging overgrazing and allowing time for rest and growth of vegetation.

ATTRA National Sustainable Agriculture Information Service has a helpful publications for educating farmers regarding organic pastures at *attra.ncat.org/livestock-and-pasture/*. This publication discusses all issues relevant to organic pasture for goats and other livestock to be certified organic, including fence construction.

Feed and Supplements. Organic livestock must be fed only hay, grain, milk replacer, minerals and any other supplement such as kelp or beet pulp that is certified organic. This means that it may not be genetically modified and may not contain synthetic hormones, antibiotics, coccidiostats or other restricted materials (see sidebar). Goats may not be given any of those additives directly, either. The bedding used for goats must also be certified organic, whether it is straw, wood chips or wood pellets, because goats may eat the bedding.

Restricted materials for organic livestock:

- animal drugs and synthetic hormones
- plastic pellets
- urea
- manure (including poultry litter)
- slaughter by-products
- excessive amounts of feed supplements or additives
- synthetic amino acids

An exception to the 100% organic requirement is for a dairy goat herd that is being converted to organic management. Those goats may be fed up to 20% conventional feeds for the first nine months of the transition, but then must receive 100% organic feeds after that time.

One major impact for farmers who are used to bottle-feeding their dairy (or other) kids with milk replacer is the requirement that milk replacer be used only on an emergency basis. This is for two reasons: 1. Dam-raising is more natural and sustainable, and 2. no organic milk replacer currently exists for goat kids, although in the UK one was introduced in 2008 for calves.

The area of the country, as well as the amount of land a goat farmer has available, can be an impediment to completely meeting the feed requirements. Farmers with a small acreage, and in the desert, for instance, will be unable to grow their own hay, alfalfa and grain and will need to rely on what is available commercially. In some areas, finding organic feeds is next to impossible. In many cases, despite availability, the cost is prohibitive. Each farmer needs to determine what these costs will be, along with the market rate for organic meat, milk or fiber, to determine whether raising organic goats is economically feasible.

Feed mills are often willing to work with a nutritionist and create a custom feed mix using organic products, if they are available. Ideally, the feed mill will be local so that shipping costs can be limited. One way to provide custom organic goat feed for less cost is to find a group of people who are interested in purchasing as a cooperative and having large quantities made at one time. It may take some coordination and time to get the bugs worked out of a distribution system, but once a group of local, like-minded farmers have signed on and prepaid, you are ready to go.

Health Care. Health care is one of the most challenging aspects of going organic with goats. A goat cannot be certified organic if it has been treated with antibiotics, or a synthetic or nonsynthetic substance that is prohibited by the law. Yet, goats do get sick with disease that requires the use of such substances. In fact, producers *must* treat sick animals, even if doing so will cause them to lose their status as "organic." So, the challenge is, whenever possible, to find treatment methods that are organic and that work.

The Livestock Healthcare Standard requires that producers use preventive health care practices (vaccination is allowed), not treat goats that are not sick (e.g., giving antibiotics or dewormers routinely), and make sure their living conditions and feed ration promote good health. Producers who do their homework will find that some breeds (or cross-breeds) of goat are more resistant to parasites and other diseases and consider obtaining these goats for breeding stock. For example, Spanish goats, which have been feral in the southern US for many years, are more resistant to parasites and require less hoof-trimming than some other breeds. Although not dairy goats, they have the advantage of being good meat goats, as well as providing cashmere for fiber.

One of the most difficult health care problems encountered when raising goats organically is controlling parasites. Ivermectin is the only chemical dewormer is that is allowed for use on organic goats. However, it may only be used if the goat is determined to have a parasite overload based on fecal egg counts. It also may not be administered in breeding stock during the last third of gestation or when they are nursing kids that are to be sold, labeled or represented as organically produced. When used in dairy stock, the dewormer may not be given to goats for at least 90 days before milk production or the production of organic milk products.

Besides prohibitions on the times dewormers may be administered, another problem exists: In some areas, parasites have become resistant to Ivermectin, so even when goats are treated, they may still have problems. Some producers use herbal dewormers, but these have not been shown in controlled studies to be effective. Fortunately, alternatives to chemical dewormers is an area in which a lot of research is now occurring, so perhaps in the future this will be less of a problem in keeping goats organically certified.

The health standard also requires that physical alterations— such as tattooing or other identification, disbudding and castration—be done only in a way that promotes the welfare of the goat and minimizes pain and stress. Although there are no hard and fast rules, these are areas that must be addressed in an Organic System Plan and farmers need to be able to show that the way they perform these procedures meets the criteria.

Organic goat farmers need to educate themselves and make sure that their veterinarian is aware of organic standards in regard to medications that are often recommended or prescribed for goats. Consider giving them a copy of the regulations or the *Livestock Workbook* if they are not already familiar with the program. That way you can work together to determine how best to treat your goats when they do get sick and not mistakenly give a prohibited drug.

The *Journal of Ethnobiology and Ethnomedicine* has produced an excellent reference on the use and effectiveness of medicinal plants for goats and other ruminants. This document can be found at ethnobiomed.biomedcentral.com/articles/10.1186/1746-4269-3-11.

Other Considerations. The National Organic Standard contains other requirements that farmers need to take into consideration. Manure management may or may not be a problem, depending on the acreage and number of goats being kept. One major intent of this provision is to keep manure out of waterways—a problem for larger farms and less so for small farms.

Recordkeeping is critical to raising goats organically and a recordkeeping system must be set up prior to applying for organic certification. Required records include "all records of the operation," and they must be understandable and available for inspection. Some of these records include identification for each goat including whether it was born on the farm or purchased, all veterinary and other health records for each goat, and feed information, which includes keeping all feed tags from feed that is purchased.

The standard also addresses requirement for processing of goat products. For example, organic meat may not come in contact with non-organic meat and no synthetic materials may be used during its processing. Farmers need to review the standard to determine whether other requirements for food processing will affect their operation and whether they can be met.

Certification Process. Obtaining an initial certification requires the following steps:

- Find a certifier in your state. (A list can be found at the USDA Agricultural Marketing Site, ams.usda.gov.
- Complete an application form. (Note: Some federal funds are available through states to reimburse applicants for obtaining organic certification)
- Describe, in writing, practices and procedures to be used.
- Make a list of each substance you will use in production, noting its composition, source, and where/how it will be used.

- Describe how the plan will be implemented and monitored.
- Describe the recordkeeping system(s) that will be used comply with requirements.
- Describe of practices and procedures to be used to ensure that organic and nonorganic products are not mixed.
- Schedule an on-site inspection by a certifier.

Initial certification is granted in perpetuity, but farmers must pay a certification fee and update their initial Organic System Plan every year—it isn't a one-time deal. Small farmers who market less than $5000 of organic products annually are not required to apply for organic certification, although they must still comply with production and handling requirements. While the milk or meat may be marked "organic," farmers who rely on this exemption may not use the organic seal on their products.

Conclusion. Farmers who think they may want to get into organic goat production should first learn everything they can about organic certification under the federal and their state laws. A good resource is the National Center for Appropriate Technology's (NCAT) *Livestock Workbook: A Guide to Sustainable and Allowed Practices* at en.calameo.com/read/000715430ba7a52daccae. This book not only includes much information on organic livestock practices, but contains a checklist that farmers can use to prepare to switch to organic systems. It can be used to determine what steps to take to begin working toward certified organic production.

The first step in making the switch is to begin transitioning pasture and cropland. Remember that it can take as much as 36 months for this step, if the land is not already certified organic. While implementing this step, farmers have time to think through and create the required Organic System Plan, talk with a certifier and start thinking through policies and procedures to be implement, as well as developing a good system for recordkeeping.

Farmers who don't have the acreage or right locale to grow feed, need to start looking for feeds that are grown organically or use organic ingredients. Although this may be an insurmountable barrier for some, there is the option of using as many certified organic ingredients as possible, and then finding "no spray" ingredients to make up the rest of the feed. Although this will prevent organic certification, it will still be a good marketing point for sales and possibly a higher price for meat and milk and can be a positive first step for potential future certified organic production.

Preparing Your Buck for Breeding Season

There is a period of time in the fall and winter when I don't even want to touch my bucks, if I can help it. This is because they are generally soaked in urine (goat cologne) and they are no longer the likeable creatures that they were during the summer. In fact, once they go into rut, they can be downright aggressive and sometimes seem to mistake me for a female goat.

To help them through this period without the normal maintenance they usually receive, I make a point of getting them ready and in prime condition by August or September. Bucks will spend most of their time in rut fighting, not eating, and staying up all night pacing along the fence line trying to get the attention of the does. This is a huge stress on them, so it is important that they don't have a high parasite load, they are in good body condition and their hooves are well-trimmed. It is also why bucks tend to live shorter lives than does or wethers.

For goat owners who rely on fresh pasture or browse to feed their goats during the spring, summer and early fall, it is time to make the shift to good quality, nutritious hay and maybe even some grain or feed that is designed for goats. I have found senior horse feed to be helpful in keeping up the nutritional status of goats who are compromised.

Adding extra nutrition will ensure that bucks don't go into rut in a run-down or thin condition—because they are guaranteed to look bad by the time it is all over (especially if there are does in the vicinity). When adding grain to a buck's ration, make sure to start slowly and gradually increase the volume. Grain can have an adverse effect on the digestive system if a lot is added at once.

Of course, you always need to have a good goat-specific mineral available free choice to ensure that they are getting the nutrition needed.

This is also a good time to do a fecal exam on each goat—whether you are using your own microscope or sending the sample to the lab. Even if their feces are pelleted and seem normal, it is good to check their parasite load to make sure there is not a problem brewing. (No goat needs to be completely free of internal parasites; the parasites should be at a low level for optimum health.) If any bucks are coughing, consider having a special fecal exam, called a Baermann, done. This will test for lungworms. Then make sure to use a dewormer that will actually kill the parasites that you want to target. Work with your vet to learn what works and does not work in your area. You also may need to repeat the dewormer a second time, often 10 days later.

Another—possible the most important—step is to trim the bucks' hooves. Although it will likely be necessary to trim them again at least one time during the breeding season, try to get hooves into ideal condition. Summer is a good time to get rid of any rot or problems that may have developed over the winter, which will occur if the buck is living in a rainy area. You will need sharp clippers, blood stop powder, in case you slip, and Hoof 'N' Heel or Koppertox to treat any rot. I recently purchased a small Dremel tool that works really well for evening the hoof pad and the sides of the hooves.

Another step is the get the bucks' skin and hair cleaned up. I routinely brush them through the summer and treat lice or mites that may have developed. I also usually give them copper oxide wires in their food after the end of the breeding season when they look really rough, and again prior to breeding. Make sure they get a good brushing to remove any flaky skin. Consider adding free choice kelp— which can help with dandruff and also contains trace elements such as selenium and iodine.

Some people suggest bathing and even shaving bucks to get them clean and to minimize the smell. Both of these strategies will also help to prevent lice, which are more common in goats with longer hair. You will need to consider the weather when shaving or clipping bucks; they will be more prone to getting sick if the weather is too cold.

Once a buck goes into full rut, it will be impossible to get him into shape—so it needs to be done before that smell is in the air. Make it a routine to prepare your bucks and does every year, for this inevitable circle of life. They will be more productive, live longer and have a better quality of life.

Basic Kidding

If you are a new goat owner, or have had your goats for a year or two and just bred them for the first time, you may be worrying about them getting through kidding safely. The main thing to remember that having a baby is perfectly normal, and most of the time, it will go just like nature planned.

Goats normally deliver their kids between 145 and 154 days. Use 150 days for estimating kidding, but keep a close eye on your doe starting at about 144 days.

What to look for. The first thing is a softening of the tail ligaments. This is 100% effective, in my experience. If you check the doe long before she reaches this stage, you will know what you are feeling. I have been mistaken in a doe with very widely spaced ligaments, but that is unusual.

Visualize a V on top of the goat's rump stretching to the tail, with the point at the tail (or a peace sign, with the long line going down the tail). These are the tail ligaments, and when they go completely mushy the doe will kid within 24 and, often, within 12 hours. This is the best sign that she is entering the first stage of labor. You can sometimes tell that this has occurred if the goat seems to have lost her ability to hold the tail up. (You can find a diagram of the ligaments in *Goat Health Care.*)

Look at the udder, which normally will begin to develop and fill. In some does this will start a month in advance, in others it happens in the last 3–4 weeks, and in other cases not until right before kidding (or rarely the day after kidding). Checking regularly toward the end of pregnancy is helpful, if the doe is agreeable. Otherwise, just look at it closely for changes. It often become taut and shiny.

You may also begin to see some discharge, and the shape of the doe's body may change as the babies begin to move into position for birth. Watch for behavior changes, such as pawing at the ground, loss of appetite, more talking, personality changes (e.g., the goat doesn't want you to leave), changing position frequently and looking uncomfortable, licking herself, breathing more heavily or grinding teeth.

Although some goats will isolate themselves, I have often observed fighting with other does, and actually had to separate the mother for safety. Each doe is different and may show different signs that she is going to kid soon.

According to David MacKenzie, in *Goat Husbandry,* as long as you can see the kid(s) as a bulge on the right side and see movement, the goat is unlikely to kid within the next 12 hours

Kidding, or parturition, is normally divided into three stages:
- First stage labor is when the uterine contractions dilate the cervix by forcing the placenta, fetus, and amniotic fluid against it. This can last up to 12 hours in first-time moms, but is often faster for those who have previously kidded. Again, every doe is different.
- Second stage of labor is the period in which the doe pushes the kid(s) out. It usually lasts less than two hours, but can be longer.
- Third stage of labor is expulsion of the placenta and the reduction of the uterus back to its normal size. In most cases the placenta is passed within an hour or two after birth, but in rare cases, it can take hours. The uterus

does not reach its pre-pregnancy size until about four weeks later.

First Stage. Birth starts with secretion of estrogen by the ovaries, which cause the uterus to contract. At this point, you will not feel the kids moving, the bulge in the doe's right side will change, and the rump will begin to slope more. This may not be visible to any but the trained eye.

You will see restlessness begin in the doe. If you have a clean kidding pen prepared, now is the time to move her there. Like all mammals, goats like a quiet, safe place to have their kids. It should be lit well enough (or have access to light) that you can see what you are doing if you need to help, but dim enough to be comfortable. It shouldn't be too small, so she can move around as the labor progresses. Avoid putting water in the pen; kids have been known to drown in it and the doe will be focusing on the task at hand. If you do want to give her water, make sure it is warm and take away any that is left.

Around this time you may see a thick discharge. This means that the doe has lost her cervical plug. You will likely see a change in discharge as labor progresses. It thickens and changes color; in some cases it may be tinged with blood. This is normal. What is not normal is thick, rusty-brown discharge, which may indicate a dead fetus. If you have questions, contact your veterinarian or an experienced goat breeder.

Your doe, at this point, will probably be repositioning herself regularly, trying to get comfortable. She may start licking herself or objects, "mama-talking" (a special talk reserved for welcoming kids), or in the case of a very spoiled goat, demand that you stay there and pet her throughout.

39

Second Stage. The second stage of labor is where the real work begins. At this point, the babies have lined up for birth and the doe begins to push them out, in sync with the uterine contractions. The contractions become stronger and closer together.

Some goats deliver standing up, others prefer lying down. The doe may or may not cry out at this point. It depends on how stoic she is.

The first sight that tells you the labor is progressing is what looks like a balloon at the opening of the vagina. This is the membrane surrounding the baby. The doe may start licking in earnest between pushes; sometimes situating herself so she can lick up the amniotic fluid. With some more pushes, you may be able to see two little hooves and a little nose, which indicates the baby is positioned properly. The kid is moving down the birth canal.

If you see just the nose and no legs, and the progress of the birth seems to have stopped, insert a thoroughly washed finger in to feel for bent back legs. You sometimes need to pull just one of these gently up to help the baby get out; in others it may take two. If you pull one leg slightly forward, it decreases the width of the shoulders and the kid should come out easily now with just another push or two.

Anytime you have to assist a goat and put your hand in her vagina, it is important to have very clean hands and short nails. Ideally, you should also wear gloves.

Often goats, especially minis, are born in breech presentation, that is, back feet first, with no problems. Frank breech position, where the hind legs are folded underneath the kids, are potentially a bigger problem, but

40

with small kids they can also be born this way. Otherwise, it will need to be corrected prior to birth, which you can do by gently pulling the feet, and then the kid, out. This prevents it from accidentally inhaling amniotic fluid and getting aspiration pneumonia or drowning.

Another presentation problem, which I have encountered only once out of hundreds of births, is crown presentation. This is where the kid's nose is pointing down toward the body, with the top of the head presenting. Because I didn't know what I was feeling, and the vet's hands were too big, we had to perform a c-section. (That kid was born four hours after his brother, and did just fine.)

Another unusual position is transverse, where the kid is sideways. This will always stop the birth and the kid has to be turned with back legs coming first and gently pulled out.

Once a kid is born, wait for the umbilical cord, if it hasn't already broken. Once the cord breaks on its own, or collapses when the blood flow stops, you can tie it off securely with dental floss in two places: an inch or two from the kid's belly and an inch past that. Only now should you cut it.

All this time mom will be licking and cleaning the baby. If the doe does not want to get up or can't reach the kid, you can bring it to her. She will continue with this behavior until the next kid is ready to be born, which can be very quickly or can take another hour or so. Longer times may be a sign of malposition, so if a placenta has not been delivered yet, and you aren't sure if there are other kids, you may want to check. Remember to err on the side of not intervening unless needed. This is where experience comes in.

There are a couple of ways to check for more kids: First, you can check inside the doe with a finger. That will at least tell you whether another kid is in the birth canal and needing positioning help.

If that tells you nothing, you can "bump" the doe. Stand behind her and with your hands on the doe's abdomen lift up quickly to feel for another kid. A more effective, but also more invasive method is to check inside the uterus with a well-washed hand and forearm that has been lubricated. I have found that having a bucket of soapy water helps this effort immensely. Wash the perineum and be gentle with your exploration. A loose-feeling uterus will contain no other babies.

I have had to do this only once in my seven years of kidding experience. In that case, the doe had ringwomb, which means the cervix will not dilate enough, and I had to slip the cervical lip around the large kid's head.

Usually you will know that doe is through kidding.

If you deliver a kid that is not breathing and seems very weak, you can try baby CPR, or simply hold tight to the kid (one hand on the leg and one on the neck to stabilize the head) and swing it back and forth in a 90 degree arc to clear the mucus. This is what I did with the c-section kid born four hours after his brother. If the kid is not able to suckle, you may need to tube feed it (see next chapter).

Third Stage. Once the kids are born, they should start nursing, which causes a release of oxytocin—also known as the bonding hormone. It not only helps mother and baby bond, but it stimulates uterine contractions that lead to delivery of the placenta and closing of the cervix. You will sometimes need to help the kids find the teats so they can nurse; in rare cases (once in my experience) the mothers will not know to nurse their young. Breeders who pull the kids at birth should milk the goat, as this has the same effect.

There is normally only one placenta for each litter, and it comes out after the birth. I understand that with some goats, more than one placenta may exist and it may rarely be expelled between deliveries of kids.

You should normally expect to see that the doe has a bag of amniotic fluid attached to some umbilical cord hanging from her vagina. The weight of the fluid helps to pull out the placenta after it detaches from the uterine wall.

Failure to deliver the placenta may indicate that another kid is still in the doe. Never pull on the membranes to remove the placenta, as that can cause ripping and lead to problems later. A placenta is not considered retained in a goat until at least 24 hours has gone by. You may obtain a prescription from a veterinarian for oxytocin for a retained placenta but do NOT routinely use it.

Do not assume that if you found already-born kids and did not find a placenta, that it is retained. Goats, like all mammals, often eat their placentas.

Once kids are born, dip their navels in 7% iodine to prevent navel ill. Make sure they are thoroughly dried, especially if the weather is inclement. They need to receive some colostrum within the first hour, if at all possible. Once mom has completed her job, I have a ritual of bringing hot oatmeal with molasses and a bucket of hot water to her. The water replenishes her system, and the oatmeal is a great treat with the added benefit of being galactogenic (helping to produce milk).

Make sure the kids have a cozy spot, with a heat lamp, if necessary, and then get the camera!

Tube Feeding a Weak Kid

Buttercup, a three-year-old Oberian doe, unexpectedly went into labor before the calendar indicated that she should. Upon checking breeding dates, I discovered that my farm partner and I had different breeding dates. She was at either 140 or 145 days gestation. That would make the kids either premature, or just on the cusp of maturity. Still, it didn't occur to me that there might be a problem with the babies.

The labor was normal and uneventful, and around 10:00 pm Buttercup delivered a little doeling, then a buckling and then a stillborn doeling. The problems began with the first kid, who was in respiratory distress and was having trouble getting a breath. Her tongue was hanging out and despite stimulation by both Buttercup and me, she was very weak. Her brother followed suit, and despite swinging and removing mucus with a bulb syringe, neither of the living kids could stand up. They were floppy, weak and had no sucking ability.

This was one of those times that I had to pull out my book *Goat Health Care* and relearn how to use a stomach tube to give them some warm colostrum. If they were mature enough, it would be the only chance they had to survive. Anyone who raises goats is wise to include a tube and syringe designed for such feeding in their birth kit to give weak or sick kids a fighting chance.

You can purchase a flexible rubber feeding tube along with a 60 ml syringe with an irrigation tip for $5 or less at most veterinary supply stores, or from a veterinarian. The tube has a tapered end, which attaches to the syringe. The cost is

45

minimal, compared to paying a veterinarian; and while it seems scary to tube feed, it really isn't that hard.

The biggest fear that people have about tube feeding is that they will accidentally get liquid into the goat's lungs. Although you need to be careful, it is much easier to get the tube into the stomach than into the lungs, and there are several ways to check to make sure it isn't in the lungs before you add milk or colostrum.

To determine how far to insert the tube, measure from the kid's nose to the center of the ear base. Then measure from the ear to the chest floor and mark the feeding tube with the sum of those two measurements. That mark is how far the tube must be inserted into the kid's mouth. (If the tube cannot be inserted that far, it is the first sign that it is in the windpipe (trachea) rather than the stomach.)

Tube Feeding Supplies

- ✓ Feeding tube, warmed with hot water to soften
- ✓ 60 ml syringe with irrigation tip
- ✓ Colostrum or milk
- ✓ Warm water
- ✓ 6 ml syringe

Although tube feeding can be done by one person, having a second person to hold the kid is better. Some kids (like Buttercup's kids) are too weak to even fight back, but others may have more spirit but yet be unable to suck.

To tube feed, hold (or have someone hold) the kid on your lap and tilt its head back slightly so the tube has a straighter path to follow. Open the kid's mouth a little by pressing on one side of the jaw with your fingers. Take the softened tube and slowly slide it down the kid's throat, small end first. If it does not go as far as the mark, slowly pull it out and start over. (I have never had this happen.)

If the kid was crying before the tube was inserted and suddenly stops during the process, slowly pull the tube out and start again.

Putting one hand on the front of the kid's throat will help you feel when the tube enters the esophagus. When the mark on the tube is at the opening of the kid's mouth, you are there.

There are several methods for checking to ensure that the tube is in the right place. The first is smelling the end of the tube for a milk smell coming from the stomach. I didn't use this method in the case of Buttercup's kids because they were just born and had no milk in their stomachs.

The second method is to place the end of the tube into a cup of water. If bubbles come out, the tube is in the lungs. I have done this, but it can be unwieldy, especially if you are working alone.

The third method is to blow gently into the tube to see whether the lungs inflate. I have not tried this method, in part because I have concerns about blowing too hard into fragile newborn lungs.

I chose the fourth method: Listen at the end of the tube for the little crackles that are the sound of breath. I heard no sounds so I was ready for the feeding.

After determining that the tube is correctly placed, you are ready to feed. Attach the 60 ml syringe to the feeding tube. Use your 6 ml syringe filled with warm water to add water to the syringe to ensure that it goes down properly and is not twisted. If everything seems fine, pour the colostrum or milk into the tube, while holding it up higher than the kid. (The plunger is not needed for this procedure, because gravity will pull the milk down.)

After the milk or colostrum is gone, add another 6 ml more water to rinse the syringe. This step is not essential, but can help to prevent milk or colostrum from going into the lungs while removing the tube, if some is left in the syringe or tube. Then withdraw the tube slowly, but in one smooth motion.

In some cases, you will see a striking difference in the kid. It may stand up within minutes and even show interest in nursing shortly afterward. In others, you may need to do several such feedings before the kid develops the necessary strength.

In the case of Buttercup's kids, their prematurity made life incompatible and the tube feeding had no effect. Their lungs were not well enough developed for them to survive without more treatment than an average goatkeeper like me can provide.

Coccidiosis: A Kid Killer

Kidding season was a great success and your goats—both moms and kids—are healthy and happy. The barn is a little more crowded than usual and it's harder to keep it clean. Then, two to five months in (around weaning time), a kid develops diarrhea, seemingly overnight. You get that under control with a little kaolin pectin or probiotics and slippery elm, and then another develops it. Soon, if the culprit is not found, most of the kids develop diarrhea. Then, the worst happens—several kids suddenly die. What now?

Assuming that the problems are caused by intestinal worms, some goatkeepers will deworm their herd. However, the thinking on that has changed over the years due to development of resistance by worms to various anthelmintics (dewormers). If you haven't done so already, it's time to get a fecal sample to find out the cause of the problem and then treat it.

For only about $100, you can get a microscope and slides to run your own fecals, and pay for it all in the first year by not purchasing the dewormers and anticoccidials that you may have been giving willy-nilly. You won't have to wait to contact a vet, or send the results to a lab for evaluation. You can even make your own flotation solution from salt or sugar. If you don't want to go that route, you can take it to your vet or send it to a veterinary lab. Contact your vet or a vet school to find out how to do this.

Once a fecal is run, you may learn that the culprit is not worms, but coccidiosis. Coccidiosis is an intestinal disease caused by a protozoan in the genus Eimeria. These one-celled creatures are host-specific, which means that goats cannot get them from chickens, dogs, horses or any other animal. (There may be some crossover in certain Eimeria species between sheep and goats.)

These critters are normally present in goats, and their environment. Only when they overpopulate and get out of control are they a problem. The protozoa attach to and destroy the lining of the intestine, as well as interacting with the digestive microflora (good bugs that help with digestion). The more oocysts (the life stage at which the protozoa are released in the feces) eaten by a goat, the more likely a problem will develop. There is no transmission through milk or in utero.

Studies have shown that when goats are heavily infected with coccidia, they are also more likely to have higher loads of other parasites, such as stomach worms. This is undoubtedly linked to the decrease in good microflora in the gut.

Coccidiosis can affect both young and old and is spread through contact with infected feces. The effects are most severe in young, old or weak animals, which lack the necessary immunity. Included in this category are does that have just kidded, and kids that are newly weaned.

Coccidiosis is also much more likely in stressed, hot or cold, overcrowded herds in unclean circumstances. In addition, it is more of a problem in wet, warm climates than in those with harsh winters or in the desert.

Coccidia are often present in the digestive system of even healthy animals. Only when they have the opportunity to overpopulate do they become a problem. When I anticipate a problem this year—for example when we have a milder than normal winter and a very rainy spring, I get my microscope out and began to check feces of various goats, so I can get a handle on any problem that may be evolving.

How is Coccidiosis Spread?

Does that are infected at kidding may contaminate the area with oocysts that are released due to stress of kidding.

Young kids that live in these areas are then at risk. Other stresses, such as moving to a new farm, feed changes or additions, overcrowding, or a drop in temperature, may be all it takes for a problem such as diarrhea to develop.

Kids are notorious for tasting things, so feeding on the ground is a good way to spread the disease. Illness can occur from 5–13 days after eating coccidia in feces. The main sign is diarrhea, sometimes with mucus or blood; dehydration; emaciation; weakness; loss of appetite; and, ultimately, death. To make diagnosis even more difficult, some goats develop constipation and die without ever getting diarrhea.

Infection with Eimeria affects the lining of the intestine, which can cause pain and blood loss. A goat that recovers may still have ulceration and scarring of the intestine— leading to stunted growth caused by malabsorption. In worst case scenarios, the goat may even develop liver failure.

A clinical diagnosis of coccidiosis is based on the number of oocysts (eggs) found in feces that are examined under the microscope. The numbers of oocysts can be phenomenal, from tens of thousands to millions per gram of feces. In kids with loss of appetite and failure to gain weight, numbers may still be high with no diarrhea. Suspect coccidiosis in thin, unthrifty goats that are not growing properly, even if you see no diarrhea.

How is it Prevented?

Because feces spread coccidia, strict sanitation is important. Some breeders regularly use a prevention program to avoid coccidiosis altogether. This involves using a coccidiostats such as amprolium, decoquinate, or lasalocid. These products can be added to milk, feed, or water. This is easier if kids are raised separate from milkers, to avoid contaminating the milk supply.

Talk to your veterinarian to find out what he or she recommends for your particular situation. Make sure that you follow the milk withdrawal and meat withholding requirements for the drug that is chosen.

Some other suggestions for avoiding problems include:

- Clean kidding pens between does.
- Keep kid pens or other areas as clean and dry as possible.
- Make sure to change food and water that may have become contaminated with feces.
- Cover hay or mineral feeders or mineral blocks that kids might be likely to jump on.
- Clip does' udders prior to kidding if kids will be nursing.
- Never feed goats on the ground.
- Control flies, which can carry coccidia from place to place.
- If you are bottle-raising kids, consider separating from the adults in clean pens.
- Muck your barn frequently, or remove manure as much as possible.

How is it Treated?

Treat early to reduce the severity of the disease process. Sulfa drugs, such as sulfaquinoxaline and sulfadimethoxine (Albon), and amprolium (Corid), available over-the-counter, are used to treat coccidiosis. *Merck Manual* states that amprolium has poor activity against certain species of Eimeria, so it may not be the best choice. In addition, it can lead to thiamine deficiency (also known as polioencephalomalacia)—so injections of thiamine or fortified vitamin B may also be required.

Treatment with these two classes of drugs is usually five days long, as an oral drench. Make sure that a kid with coccidiosis is well-hydrated because diarrhea can lead to dehydration. Treat for the full course even if the kid improves in the first few days.

Some veterinarians now recommend a drug called toltrazuril, which only has to be given one time, and works on the whole life span of the protozoa. This is in contrast to amprolium and monensin—which are effective during the early stages, and sulfa drugs—which are effective in later stages. The dose for goats is two times that for sheep or cattle.

Other Thoughts

Some goat breeders use the "wet tail" method for determining when to treat for coccidiosis. With this method, whenever a kid (particularly post-weaning) has a tail that indicates loose, watery stool, they treat. One of the reasons I like the sulfa drugs for treatment is that they are also effective against some bacterial diarrheas.

Ideally, goat owners will do a fecal exam as soon as they notice the slightest problem, so they can determine what organism—if any—is causing the problem. Other possible causes include giardia, enterotoxemia, salmonella, and many others.

One option is to treat with an antidiarrheal product such as Pepto-Bismol or kaolin pectin when a goat develops diarrhea, to see whether you can get it under control without the use of harsh medicines and while awaiting fecal exam results.

For those who prefer herbal treatments and prevention, tannin-containing plants—such as pine needles and oak leaves—were found in a Korean study to decrease coccidia egg counts.

Interestingly, having spotless pens may lead to more clinical coccidiosis in kids, because they need some exposure to become immunized to the coccidia. Finally, overuse of anticoccidial medications can, as with other intestinal parasites, lead to resistance and eventually they will not work.

Anyone who is raising more than a few goats—especially if they are kidding, showing, or otherwise exposed to stressors—may eventually have to deal with coccidiosis. Being prepared, knowing what to expect and acting quickly can keep those goats healthy and save lives.

Urinary Calculi in Wethers

My brother-in-law called one cold winter morning to ask if I could tell him what was wrong with one of their goats. He was standing hunched up, straining and not eating or drinking. Immediately I knew that he had urinary calculi and that things looked bleak for the goat.

They took him to their vet, where they decided to have him euthanized. Although the initial treatment may be successful, in my experience, if a wether develops stones one time, he will continue to do so. Few goat owners are prepared to spend thousands of dollars keeping a wether alive, particularly if its only job is eating brush.

As I later learned, my pre-teen nephew had been responsible for watering the goats. The freezing temperatures required that buckets be refilled many times a day and he had not been diligent in doing so. It was a hard lesson.

If you keep wethers as pets or for brush-eating, be aware that some of them are prone to developing urinary stones (also called calculi). Calculi are caused by an accumulation of minerals or other compounds that can cause trauma to the urinary tract and bladder and obstruct the flow of urine out of the body.

One reason that calculi are more common in male goats than in females is because of their anatomy—the urethra is a common site for blockage.

Several factors can cause wethers to be more prone to calculi. These include genetics, early castration, diet and lack of water. In terms of genetics, it is hard to know whether a specific wether has inherited a problem with this, because many male goats are used for meat or not registered. Most goat breeders do not keep wethers as pets and they also don't follow up on where the goats went or how they are doing. They may have been sold at auction, or re-sold several times before they develop a problem.

Age at Castration. Wethers that were castrated early are more likely to develop problems with urinary stones. When purchasing a wether as a pet or brush-eater, ask the seller when the kid was castrated. Ideally, in order to allow the urethra to mature, it is best to wait until at least two months of age to castrate a buckling.

Although I am unaware of a similar study on goats and urinary calculi, a study done in cattle showed that although bulls and steers can both get calculi, bulls are more likely to pass a stone that would obstruct a steer's urethra. This is because the testosterone produced by the bull makes the diameter of the urethra 25% larger than that of a steer. The same is most likely true for bucks and wethers.

Water Intake. The most important factor in preventing urinary calculi is to increase water consumption. Concentrated urine can lead to urinary stones. Keep water bowls clean and fill frequently with fresh water. During the winter, use heaters or plug-in buckets or give them hot water regularly throughout the day to encourage consumption. During the summer, make sure the water is in a shady spot.

Some goat owners have found it effective to give 1/3 cup apple cider vinegar diluted in a cup of water twice a day. For wethers that don't seem to be consuming enough water, you can try adding a sugar-free drink mix to the water. This will make the water taste better, so they will drink more.

Diet. Diet is another key to preventing urinary stones. Wethers should be fed grass hay as their main source of food, along with whatever browse they have access to where they live. Alfalfa and other legume hays have more calcium than is healthful for them—and they may develop calcium stones.

Wethers should also not be fed large quantities of grain, as it can lead to phosphate stones. I recommend not giving wethers any grain at all, to be on the safe side. Once they are grown—and even while they are growing—they can get all their nutrition from good hay and minerals.

Free-choice or frequent feeding is also considered to be a factor in limiting urinary stones, because of chemical reactions that affect urine concentration at each feeding. So make sure your wethers have hay available, or give them small portions several times a day

Importance of Salt. Increasing the salt concentration in the diet increases water intake, leading to more diluted urine. According to studies in Canada, free-choice loose or lick salt is not effective in preventing stones in male goats that were considered at risk for silica calculi.

They found that mixing the salt directly into the feed was the most effective means of providing it to the animals. (Of course, this is problematic if you are not feeding your wethers grain.) A friend of mine has also had success with spraying hay lightly with a salt-water solution. Another trick is to give some salted corn chips as a treat.

Ammonium Chloride. Many goat owners add ammonium chloride to the diet of bucks and wethers. It should be given at a rate of 1% of the dry matter in the diet. This reduces the pH of the urine (makes it more acidic), which makes various types of urinary stones more soluble in the urine.

I tried adding ammonium chloride to my bucks' food at one time, but they found it very distasteful and weren't interested. This is where a sweetened drink would also be helpful.

One study showed that ammonium chloride given over a long period of time reduced the mineral content in the bones of ewes. It may have the same adverse effect on a wether.

Urinary calculi affect only some goats. Be aware of your goats' health and make sure that they have a proper diet, water and salt. These steps will go a long way toward preventing urinary stones and keeping your goats alive.

Pneumonia in Goats

If you raise goats, sooner or later you will have to deal with pneumonia. It is a common illness in goats, which can occur in both kids and adults. In some cases, pneumonia is infectious, but in others, it is specific to the goat and not transmissible. Pneumonia is an inflammation of the lungs, caused by some sort of irritant that affects the lining of the lungs.

Causes of Pneumonia

- ✓ Sudden changes in weather

- ✓ Viruses, including CAEV or

- ✓ Poor nutrition or dietary changes

- ✓ Stress of transport

- ✓ Microbes, such as *Mycoplasma*

- ✓ Bacteria, such as *Pasteurella*

- ✓ Lungworms

How do I tell if my goat has pneumonia?

Signs that a goat has pneumonia include a moist painful cough, difficulty breathing, runny nose and/or eyes, loss of appetite and depression. Not all coughing or runny noses are caused by pneumonia, however. The first step when you notice that a goat is acting "off" or has some of the symptoms of pneumonia is to take a temperature. A temperature between 104° and 107° F, points to pneumonia.

Next, put your head down to the goat's side, back a little from the shoulder and listen to the lungs. (If you have a stethoscope, use that.) If your goat has pneumonia, you may be able to hear crackles when she breathes.

What causes pneumonia?

The most common cause of pneumonia is bacterial infection. Mammals normally have bacteria in their lungs. It is only when they have other stressors that they are unable to stop an overgrowth. In goats, the most common bacteria that cause pneumonia are *Mannheimia haemolytica* and *Pasteurella motocida.*

Viruses are common in goats and can sometimes increase susceptibility to respiratory infections by inflaming the throat and lungs. The lentiviruses, which cause ovine progressive pneumonia (OPPV) and caprine arthritis encephalitis (CAEV), can lead to a chronic pneumonia in goats. Unfortunately, these cannot be cured and will eventually debilitate and kill a goat.

Lungworms can also chronically irritate the lungs, leading to pneumonia. They are most common in the cooler months, and are spread when larvae or coughed up, or when they are released in the feces and then mature. They are killed by hot water and hard freezes, but because they use snails and slugs in their development cycle, they thrive in cool, wet conditions.

Lungworms cause a dry cough, which is easily treated, although it may leave the goat with a chronic cough if they are severe enough. A special fecal analysis, called a

60

Baermann, must be done to definitively diagnose lungworms. Treatment is 2X the cattle dose of Valbazen for three consecutive days, or 2X the cattle dose of Ivermectin for three consecutive days. Repeat in 14 days.

Mycoplasma is a particularly frightening cause of pneumonia. It is not a true virus and is not easily cured with antibiotics. It can cause a contagious caprine pleuropneumonia in goats. It also causes some of the same symptoms as CAEV, including inflammation of the joints, the udder and the eye. The drug of choice for treating mycoplasma-related pneumonia is tylosin, or Tylan. Mycoplasma pneumonia is highly contagious and can be spread to kids through milk. (Read the excellent article on Mycoplasma by Gianaclis Caldwell at gianacliscaldwell .com/tag/mycoplasma/.)

How can pneumonia be prevented?

As with many health issues in goats, management is the key to preventing pneumonia. Make sure that goats have proper shelter and can get out of the elements when they need to. Avoid overcrowding and other stressors. Make sure they have a good diet, with adequate minerals and fresh water. When a goat does get sick, isolate her from others. This serves two purposes: It keeps the other goats safe from getting the illness and it keeps the ill goat from getting bullied by others when he is down. (A caveat here: if you only have a few goats and they have never been separated, you may determine that it is too stressful on the sick goat to separate her from her family.)

Goatkeepers with larger herds sometimes vaccinate their goats against bacteria that cause pneumonia, but in small herds it is uncommon enough that the cost and trouble may not be justified. I have never vaccinated my goats against pneumonia and have not found it to be a problem. If you are interested, Colorado Serum Company makes a vaccine for this. The dosage is 2 ml SQ twice, two weeks apart.

Is pneumonia in young kids different?

While newborn kids are exposed to the bacteria by their mothers and other goats, they also receive antibodies in colostrum, which helps to keep them healthy. This is one reason it is essential for newborns to receive goat colostrum. Other factors that can lead to pneumonia include living conditions—overcrowding, lack of ventilation and lack of cleanliness; stress—travel, being suddenly removed from their mother, or even chased by children; or another condition that has injured the lungs—such as lungworms or a virus.

Young kids are of particular concern because they can die quickly once they get pneumonia. They tend to lose weight, become lethargic, develop a runny nose and rapid breathing and run a moderate fever. The first sign is their loss of interest in nursing. This is easier to detect in bottle-fed kids for obvious reasons.

Time is of the essence in treating pneumonia in kids because if it isn't detected and treated early, their lungs may get damaged. When this happens the kid may recover, but will be prone to further bouts of pneumonia, development of a chronic cough, and may not grow as well as healthy kids.

If you find that a kid with other symptoms of pneumonia has a high fever, you can give him ¼-1/2 of a baby aspirin. Because a fever is nature's way of fighting infection, a moderately high temperature may not need to be decreased by medicine.

How is pneumonia treated?

Regardless of the cause of pneumonia, giving the goat plenty of fluids is essential. Especially in colder weather, adult goats prefer warm or hot water. Some people add a little blackstrap molasses or even concentrated grape juice to encourage them to drink.

Kids that are not nursing or taking a bottle need to be tube-fed. All goat owners who are planning to breed their goats should make sure to have tube-feeding equipment on hand. You never know when you will need it.

In severe cases, SQ injections of sterile water will help, unless you can afford to contact a vet to place an IV for sterile water. Give Banamine or another anti-inflammatory drug to reduce fever and inflammation and Benadryl syrup (1 tsp for kids) or another antihistamine for congestion.

For bacterial pneumonia, give antibiotics. Naxcel is labeled for goats and is considered the drug of choice for the most common bacterial pneumonias. It is a prescription drug with a short shelf life. Make sure to give the whole course of antibiotics, according to veterinarian instructions. For particularly serious outbreaks, some goat owners treat all exposed goats with a course of antibiotics to prevent more sickness.

If you manage your goats right, you can avoid most illness. But every so often one of them will get sick, no matter what you do. Pneumonia is one of the culprits. This is why it is essential to look at your goats at least twice a day to make sure they are acting like themselves. If you catch it early, the odds are good that you can nip it in the bud.

Marketing Your Goats

Learning about goat care and keeping up with what seems to be never-ending work are critical to successful goatkeeping. In addition to that, marketing knowledge is essential. However, unless you have prior experience, success in marketing your goats can be elusive. For example, I once ran an ad for four weeks in the regional farm newspaper to sell my excess wethers and had no responses.

Marketing Information and Assistance. State extension services often have marketing information that can be obtained free or for a nominal charge. If you are online, check to see if they have information on a website. If not, call and ask about hard copy publications.

One caveat I would mention is that, whether you are marketing on the internet or in hard copy publications, tracking where your customers originated is essential to determining the effectiveness of your effort. If your advertising budget is limited, why squander it on advertising that doesn't work?

Marketing information specifically related to dairy goats is hard to come by online. But info on marketing in general, particularly internet marketing, is plentiful. You can also check out your competitors' sites for ideas. Keep your website content short and uncomplicated.

Advertising on the Web. E-commerce is the latest buzzword. Selling on the internet provides you with a worldwide potential market. But want to build a website yet, there are other ways to advertise your goats on the web. Magazines such as *Goat Journal* (backyardgoats.iamcountryside.com/) or *Goat Rancher* (goatrancher.com/) have directories and online advertising. Facebook has many groups related to livestock and goats, and some are specific to certain breeds or are just for advertising.

Fairs and Other Events. County and state fairs provide another good opportunity for marketing your goats. While just being in the public eye is helpful, you can enhance your visibility with a good banner, or even a colorful display. You can include lots of pictures, a poster with the uses of goat milk, and even some samples of products. Of course, the best seller will be those cute kids performing in the pen.

Fairs are also a good place to talk with the public, hand out business cards, and educate about the breed or goats in general. Other venues for this kind of marketing include petting zoos—which sometimes need various animals—local parades, and schools. I sold soap made with milk from my Nigerian Dwarf goats, so when I attended craft fairs I brought along a supply of cards, some pictures, and my sign. I always got a number of inquiries from people who were thinking of getting some goats.

Signs. If you make a sign for the fair and other events, get an all-weather one that you can also hang on the barn during the off season. If you don't live too far off the beaten track, a sign in front of your property, giving people notice that you sell goats, is a good investment. The only downside is that you may get unexpected visitors; so make sure to include your phone number for those who are courteous enough to check before stopping by.

Share your Expertise. Besides obtaining marketing information from your county extension office, you can also be a provider of information. Find out when educational sessions are held and offer to provide one on goat raising, holistic goat medicine, or some other related subject where you have gained expertise from your experience in breeding and raising goats. This will bring out those who are new to or just interested in goat raising, and possibly a customer or two. You can also offer to speak at school functions, on local radio or at any of a number of other venues. When speaking, be sure to bring along brochures, business cards or other printed materials that the audience can take home. This way you can market your herd in both printed and oral form. People are much more likely to remember, or share with friends, when the information is in written form.

Write about Goats. While you may be just preaching to the choir when writing for newsletters and other publications targeted at those who already own goats, it will at least give you name recognition among breeders. But you can also write for other audiences, in agricultural publications, homesteading magazines, and the like. Letters to the Editor are also a good avenue for publicizing the breed and catching the eye of potential buyers. You can either respond to another article or just write an informational letter. Just remember to check online for requirements and keep it short, as letters to the editor are edited, and you will want to make sure your most important point doesn't end up cut.

I wrote a letter in response to an article on dairy goats that failed to mention Nigerians. I commented on the excellent article and then went on to describe the goats and our regional breeders' association, giving contact information.

Advertise in Publications. As you know, either by personal experience or the introduction to this article, advertising does not guarantee buyers. That is why selection of where to advertise is so important, particularly if you have a limited budget. Do you advertise in publications such as *Goat Journal*, which have a broader readership such as a general goat, an exotic animal magazine that covers many animals, or the local newspaper, or all of them?

To answer that, you need to consider who you want to reach. While the goat- or breed-specific publications will put you in touch with other breeders who may want to purchase some of your top animals, they may not reach those who are new to goats or just thinking about getting started. A mix of the two is probably the best, since advertising in the breed-specific publications will help support them and the valuable service they provide in marketing and educating regarding the breed generally.

You can find a list of goat-related magazines by simply searching the web. I would also recommend talking to other breeders to find out where they advertise, and why.

These are only a few of the possibilities for marketing your goats. There are probably as many or more that I missed. The most informal and possibly the easiest way to market your goats is simply talking to people you meet. You never know when a casual conversation in a grocery checkout line might lead to another convert to these wonderful, friendly animals, and possibly, a sale.

A Well-Stocked First Aid Kit

Supplies and Equipment

Drenching syringe

Alcohol and cotton, or alcohol prep pads—for cleaning wounds, clippers or other things.

Clippers—for clipping around wounds or abscesses

Thermometer—number one item for gauging whether you have a problem

Needles and syringes 20 gauge, ¾" needle, 3 and 6 ml—the most common sizes needed for most routine injections

Nitrile gloves—essential when dealing with potential infectious disease

Sharp, sterile scalpel—needed for draining an abscess (make sure to wear gloves!)

Sterile Scissors—for cutting umbilical cords, bandages

Clean towels (I cut mine in half)—you can use old ones or buy at garage sales or thrift stores

Elastic bandages

Sterile Water

OTC Medications and Treatments

Active Manuka Honey—licensed in Australia and New Zealand as a medicine, this can be used topically instead of antibiotic ointment. It tastes good and attracts flies so may not always be practical.

Baking soda—some people give it free choice. Good for stomach upsets/bloat and goats like it.

Betadine Surgical Scrub—this is a good thing to have for cleansing wounds

BioMycin—this version of oxytetracycline stings less than LA-200

Blood stop powder—for those times you overtrim a hoof, or a goat loses a horn

Epinephrine—I have never had to use it, but this is critical if an animal has an adverse reaction to a drug that was administered.

Dewormers—which one you have on hand will depend upon whether resistance has developed for the one(s) you previously used. I normally have Valbazen and ivermectin.

Lactated ringers—Can be used to give subcutaneous water injections if an animal is debilitated and/or has severe diarrhea if a vet cannot be reached to give IV

NFZ puffer— I got this powdered nitrafurazone from my vet after I saw him use it on the head of a goat he had disbudded, on a broken leg before it was wrapped, and other injuries.

Molasses—I use this in water after a birth to give the mom a little boost. It can also be used to get sick goats to drink.

Koppertox or Hoof and Heel—for treating or soaking feet when a goat may have foot rot

Wound-Kote—often used after disbudding and on other wounds.

Di-Methox 40% --for treatment of coccidia and other bacterial infections of the intestine.

Electrolytes—for goats that are sick, have loss of appetite or just need a boost

Fortified Vitamin B complex—can be given as an appetite-stimulant and for debilitated animals..

Sheep and goat Nutridrench—good boost for tired or weak animals

Today—an intramammary treatment for mastitis during lactation.

Pepto bismol—first-line treatment for severe diarrhea

Probiotics–to help jump-start a rumen in a sick goat.

Propylene Glycol

Penicillin

Terramycin eye ointment—goats seem prone to eye problems, so this is essential to have on hand before things worsen

Toxiban or other activated charcoal, for poisoning

Triple Antibiotic—for wounds or other injuries

Vaseline—used as a barrier for urine in bucks and can also be used to smother mites

Zinc Oxide—good for bucks who are peeing on themselves, as well as wethers that have a urethrostomy due to urinary calculi

Veterinary Medications

BoSe–I give this to my does a month before kidding. It can help with making sure the uterine muscles work well in labor

C&D antitoxin—Hopefully, this is rarely used. I have had some on hand, but found that it expired. It is critical when needed, though.

Banamine—the go-to painkiller for goats. It should be used in consultation with your vet and not long-term because it can be very hard on the kidneys. Some vets recommend that goats have the oral variety because of with injection sites, but I have not had problems.

Kidding Supplies

Use a Rubbermaid box to store all of these items so they are ready when it's kidding time.

Baby Monitor—monitor those expectant moms without being in the barn

Lubricant—essential for assisting with malpresentations or other kidding problems

Bulb syringe—for breech kids or others with breathing problems

Dental floss—if the cord does not break on its own or drips blood, use to tie in two places and cut between

Towels—great for helping to dry off newborn kids

Sterile scissors—to cut off or shorten umbilical cord

Pritchard Nipples and Plastic Pop Bottles—in case you have to bottle feed colostrum

7% Iodine—to dip umbilical cords

Plastic film can or another container with a tight lid—very handy for storing iodine and dipping cords

Bonus Chapter: The Story of Comet

Comet was a Nigerian dwarf goat who left a positive mark on this world. He was an ambassador for my farm, loved everyone, and made a lot of people happy over a few days.

When Comet was born he didn't look like he was long for this world. He was one of triplets, and the runt. He had trouble breathing and required warming and swinging to get him going. If goats had Apgar scores, he was probably a 4 at birth and then an 8 or 9 at five minutes. I took him in the house after making sure that the mom and the others were all right.

The little gold-and-white goat came around pretty quickly after getting his first colostrum from a plastic pop bottle topped with a Pritchard teat nipple. I spent that first night waking up every 2-3 hours to feed him warmed goat milk.

Then it was time to get ready for the 2001 Oregon State Fair. Comet would be going as an exhibit. Because he had no mother—other than me—he couldn't stay with the other goats. No one would protect him and he had already missed the critical time to learn about nursing from a mama, if she would even have him.

Back then, the State Fair administrators allowed to sleep with our animals. I took along an extra cage so that Comet could share the pen where I slept. He could use that for naps and to sleep in at night. I also took him a leash and collar. He had spent the first week of his life learning to walk on a leash and he loved it!

I knew something was wrong with this little goat, but I wasn't quite sure what. His tininess had me worried. He seemed to be thriving on his four times a day diet of warm goat milk, though.

After a morning of packing, herding goats into the van, and double-checking that we had everything we needed—buckets, clippers, milk stand, hoof trimmers, hay, grain, collars, leads, first aid kit, cot, sleeping bag, broom, food—we headed for the fair. The day was sunny—the last gasp of summer—but my mood was upbeat. This was the highlight of show season, with the last show of the year. It was also a time to renew friendships with other goat people and buy new goats for breeding.

We got there and then, in a flurry of activity, unloaded goats, van, and equipment. I had to put straw in pens, determine which goats go in which pens, rearrange, move them in, haul water, set up feeders with hay, and set up tack and sleeping pens. Once all of this was done, it was time to start socializing with old friends, despite being so tired I could drop.

Comet's job, most of the time, was to sit in his pen and look cute. When I got tired of asking questions or hanging out with other goat people, I would take Comet out walking. He was probably all of two pounds, just a little bit of guy, but he walked proudly on his leash. And we were the center of attention.

"Look at the cute little dog! Omigosh, that isn't a dog, it's a goat! Can I pet him?"

"What's his name?

"How old is he?"

"What kind of goat is it?"

"Can I hold him?"

Over the five days at the fair, Comet probably met several hundred people. I never got tired of seeing their eyes light up when they saw him. Little girls flocked to see him, elbowing each other out of the way to be the first to hold him. He had his picture taken with countless people of all ages. Comet was a star!

On the last day of the fair, Comet's energy started to wane. As the day wore on, he became warmer and started having trouble breathing. I consulted some goat people about what to do, conferred with the fair veterinarian and started him on antibiotics. It looked like pneumonia, which can kill kids fast. Comet continued to go downhill.

I panicked and asked the goat superintendent if we could leave early. My request was granted and we loaded up the van early, avoiding the big rush. We drove home, unloaded goats and supplies, and got Comet settled in a carrier in the kitchen. It was 8:00 pm and after dark. I was exhausted.

I gave Comet another antibiotic injection, while he cried out in pain. I felt so helpless. It was a holiday night and no vets were available. I didn't know what to do. I put a little t-shirt on him and made sure he was warm. I gave him a bottle of warm water with electrolytes, but he seemed indifferent.

I went to bed around 10:00; I needed some sleep and he did, too. I could only hope that he would make it.

At 2:00 am, I awoke with a start. *Comet! I hope he's okay!* I walked out to the kitchen, wearing only a t-shirt and looked into the cage. Comet's breathing was even more labored than before. He had mucus coming out of his nose. I held his little body next to mine and gently wiped his nose. "Oh, Comet, I love you." I told him. I felt crushed; I had reached the end of my abilities when it came to veterinary care. This was only my third year raising goats, and I had no veterinarian to call, even if I could afford to save a little runt goat.

For the next hour I carried him around the kitchen, his hot little body curled in my arms. "Comet, please don't die," I pleaded. I didn't want to give him another shot of antibiotics because I could see how much it hurt the little guy. I didn't think it would work anyway, at this point.

My heart breaking, I put Comet back into his little cage and went to bed. I knew I was losing him and I couldn't face it. I felt like a failure.

When I got up the next morning, the cage was gone and Comet was gone. I looked at my roommate.

"I knew he was going to die, so I went to bed. I couldn't deal with it." I told him.

"I know, honey," he said. "It's okay, I buried him for you. There was nothing you could do. He just wasn't meant to live."

More than 20 years and hundreds of goats later, I still remember and miss little Comet. He brought more joy to more people in his short time on this earth than many of us do in a lifetime.